The Art of War

Sun Tzu

Translated from the Chinese
by
Lionel Giles

DOVER PUBLICATIONS, INC.
Mineola, New York

Bibliographical Note

This Dover edition, first published in 2002, is an unabridged republication of the edition originally published in 1944 by The Military Service Publishing Company, Harrisburg, Pennsylvania. The English translation of the text of *The Art of War* was made by Lionel Giles and was first published in 1910 by Luzac & Co., London. Summaries of some of the translator's annotations are included in brackets in this edition. The Introduction and the chapter on Chinese Warfare were written by Brig. Gen. Thomas R. Phillips, U.S. Army, for the 1944 edition.

In this edition, there are no pages 1–4; the title page is page 5.

Library of Congress Cataloging-in-Publication Data

Sunzi, 6th cent. B.C.
[Sunzi bing fa. English]
The art of war / Sun-tzu.
p. cm.
Originally published: Harrisburg, Pa. : Military Service Pub. Co., 1944.
ISBN 0-486-42557-6 (pbk.)
1. Military art and science—Early works to 1800. I. Title.

U101 .S949 2002
355.02—dc21

2002067294

Manufactured in the United States of America
Dover Publications, Inc., 31 East 2nd Street, Mineola, N.Y. 11501

CONTENTS

INTRODUCTION

WRITTEN about 500 B.C., *The Art of War* by Sun Tzu is the oldest military treatise in the world. Highly compressed, it is devoted to principles and still retains much of its original authoritative merit. To the military student able to adapt its principles to modern warfare, it even now, twenty-five centuries after its preparation, is a valuable guide for the conduct of war. Although the chariot has gone and weapons have changed, this ancient master holds his own, since he deals with fundamentals, with the influence of politics and human nature on military operations. He shows in a striking way how unchanging these principles are.

Sun Tzü Wu, according to Ssü-ma Ch'ien, was a native of the Ch'i state. His *Art of War* brought him to the notice of Ho Lu, King of Wu [in middle-eastern China, west of Shanghai. The Capital was the present city of Wuchang]. Ho Lu said to him: "I have carefully perused your thirteen chapters. May I submit your theory to a slight test?"

Test of His Principles

Sun Tzu replied: "You may." Ho Lu asked: "May the test be applied to women?" The answer was again in the affirmative.

So arrangements were made to bring 180 women from the palace. Sun Tzu divided them into two companies and placed one of the King's favorite concubines at the head of each.

He then bade them all take spears in their hands, and addressed them thus: "I presume you know the difference between front and back, right hand and left hand?" The girls replied: "Yes." Sun Tzu went on: "When I say 'Eyes front,' you must look straight ahead. When I say 'Left turn' you must face towards your left hand. When I say 'About turn,' you must face right around towards the back."

The words of command having been thus explained, he gave them halberds and battle-axes in order to begin the drill. Then, to the sound of drums, he gave the order, "Right turn!" But the girls only burst out laughing. Sun Tzu said: "If the words of command are not clear and distinct, if orders are not thoroughly understood, then the general is to blame."

Then he started again, and this time gave the order, "Left turn!" Whereupon the girls once more burst into fits of laughter. Sun Tzu said: "If the words of command are not clear and distinct, if orders are not thoroughly understood, the general is to blame. But if his orders *are* clear, and the soldiers nevertheless disobey, then it is the fault of their officers." So saying, he ordered the leaders of the two companies to be beheaded.

Wu's King was watching the scene from the top of a pavilion; and when he saw that his favorite concubines were about to be executed, he hurriedly sent down the following message: "We are now quite satisfied as to Our general's ability to handle troops. If We are bereft of these two concubines, Our meat and drink will lose their savor. It is Our wish that they shall not be beheaded."

Authority of High Command

Sun Tzu replied: "Having once received His Majesty's commission as general of His forces, there are certain commands of His Majesty which, acting in that capacity, I am unable to accept." Accordingly, he had the two women beheaded, and installed the pair next in order as leaders in their places.

When the execution was over, the drum was sounded for the drill once more. And the girls went through all the evolutions, turning to the right or to the left, marching ahead or wheeling back, kneeling or standing, with perfect accuracy and precision, not venturing to utter a sound.

Then Sun Tzu sent a messenger to the King, saying, "Your soldiers, Sire, are now properly drilled and disciplined, and ready for Your Majesty's inspection. They can be put to any use that their sovereign may desire; bid them go through fire and water, and they will not disobey." But the King replied: "Let Our general cease drilling and return to camp. As for Us, We have no wish to come down and inspect the troops."

Sun Tzu's retort was: "The King is only fond of words and cannot translate them into deeds." After that Ho Lu realized that Sun Tzu knew how to handle an army, and finally appointed him general.

In the west he defeated the Ch'u state and forced his way into Ying, the capital. To the north, he put fear into the states of Ch'i and Chin, and spread his fame abroad amongst the feudal princes. And Sun Tzu shared the might of the king.

This narrative may be apocryphal, but Sun Tzu says in his book: "There are commands of the sovereign which must not be obeyed."

Comments on the Book

Aside from this anecdote there is very little information about Sun Tzu. His existence has been the subject of a great deal of scholarly quibbling, but the consensus leaves no doubt of his actually having lived, despite the scarcity of information about him.

Sun Tzu can boast an exceptionally long and distinguished roll of commentators, who would do honor to any classic. The first commentator appears to be Ts'ao Ts'ao, [A.D. 155-220], one of the greatest military geniuses the world has ever seen. His notes, characterized by Giles as "models of austere brevity," are so compressed that "they are scarcely intelligible and stand no less in need of a commentary than the text itself." Almost a thousand years later, Tu Mu [803-852] said that all the military triumphs and disasters since Sun Tzu's death would, upon examination, be found to uphold and corroborate in every particular the maxims in *The Art of War*.

Sun Tzu has exercised a potent fascination over

the minds of some of China's greatest men, not only military men, but, more remarkably, over purely literary men. One critic, Cheng Hou in his *Impartial Judgments in the Garden of Literature*, quoted by Giles, says that *The Art of War* is "not only the staple and base of all military men's training, but also compels the most careful attention of scholars and men of letters. His essays are terse yet elegant, simple yet profound, perspicuous and eminently practical."

European Approval

Giles says that "the few Europeans who have yet had an opportunity of acquainting themselves with Sun Tzu are not behindhanded in their praise." He quotes a letter from Field Marshal Lord Roberts of Kandahar [1832-1914]: "Many of Sun Tzu Wu's maxims are perfectly applicable to the present day. The people of this country would do well to take to heart" the following: The art of war teaches us to rely not on the likelihood of the enemy's not coming, but on our own readiness to receive him; not on the chance of his not attacking, but rather on the fact that we have made our position unassailable.

No page of Sun Tzu's book can be read without finding the distilled wisdom of a great soldier written with the aphoristic distinctness characteristic of Chinese literature.

IN time of peace, the very low pay of the Chinese soldier attracted only the peasant and laboring classes. In war a better class was attracted by higher pay and the prospect of plunder or prize money. About 170 A.D., the custom was started of raising forces from among those Tartars who had become Chinese subjects, for special service on the frontier. This was found to furnish soldiers of ability, but of doubtful loyalty.*

It is probable that for centuries most of the Chinese soldiers were infantry. About 117 B.C., a force of about 150,000 cavalry was raised for service against the Tartars, but this seems to have been exceptional. Every force had some bowmen and spearmen. In the early wars chariots were used.

Reform in the Army

Under the Emperor Taitsong (763-780 A.D.) there was considerable reform in the army. He organized his army of 900,000 men into 895 regiments of about 1,000 men each. Of these, 261 regiments were used for service on the border and 634 for service in the interior of the country.

The Chinese troops had no training corresponding to our drill or combat exercises. For instruction in horse-back archery, a shallow trench about a hundred yards long was dug so that the rider

* Much of the historical data and comment which follows may be found in General Mitchell's *Outline of the World's Military History;* the Military Service Publishing Company, Harrisburg, Pa.

would not have to guide his horse. As the horse ran down the trench, the rider shot an arrow at a target twenty or thirty yards away, but seldom hit it. Other exercises, such as wrestling, throwing large stones, and use of heavy swords seem to have been intended to encourage muscular development.

Sparse mention of the staff is found in Chinese history. Supply seems to have caused more trouble than anything else. We find one emperor who himself agreed to take charge of the supply of an army of 600,000 men so as to relieve its commander from this worry. About two hundred years later generals protested that an expedition ordered by the emperor could not be undertaken because proper supply arrangements could not be made in time. They added that supply considerations limited a campaign in cold and windy regions to a maximum of one hundred days. Supplies seem to have been drawn by slow-moving oxen, and the forage for the oxen was a large item in regions where they could not be supported by grazing.

Spies Extensively Used

For information of the enemy, the Chinese depended largely on spies. In the intervals between battles, negotiations were frequently carried on, not for the purpose of settling the difficulty without fighting, but to introduce envoys into the enemy's camp so that they could keep in touch with what was going on. Captured enemies were frequently induced by bribes, threats, or torture to disclose plans. The Chinese used advance guards, but not reconnoitering detachments. This lack of security caused them to be taken frequently by surprise.

Five kinds of spies are listed and their use is treated in detail in *The Art of War*. Sun Tzu could well understand how the German aviators were able to bomb Polish headquarters every time it was moved in September, 1939.

In the Chinese armies one arrangement in command was faulty. Mandarins who had achieved distinction as civil functionaries studied miltary tactics late in life and directed operations in time of war, while officers of experience could not expect to reach the highest grades. As might be expected, this system frequently brought disaster to Chinese arms.

Chinese Weapons

Weapons generally used were bows and arrows, spears, war chariots, swords, daggers, shields, large iron hooks, and iron headed clubs about five feet in length and weighing twelve to fifteen pounds. The standard equipment of the Wall guards consisted of a sword, a crossbow, and a shield. The bows were of three classes according to the force necessary to bend them. Each man was issued one hundred and fifty arrows. Two kinds were used and both had bronze heads. Quivers were issued to carry the arrows.

The Chinese discovered gunpowder before 255 B.C., but used it only for fireworks until much later. Isolated instances are given of its use for military purposes, in 767 A.D. and 1232 A.D., but cannon were not used for the defense of the Wall until the Ming Dynasty in 1368 A.D., about the time they were first used in Europe.

No Medical Corps

The Chinese had very little knowledge of medical science. They had even less knowledge of anatomy, because of religious convictions against mutilation of a dead body and the drawing of blood. Hence they practiced almost no surgery. They placed great reliance on diagnosis by the pulse. They claimed that by taking the pulse at different places, they could tell the state of health of the different organs, and even determine the sex of an unborn infant. As they had no knowledge of sanitation, epidemics were frequent. The most common classes of diseases were those of the eye, skin and digestion; and intermittent fevers were common. No mention is made of a medical department in the army. The losses from disease were high in the campaigns, and the wounded had slight chances of recovery.

The Great Wall

The greatest military engineering feat of the Chinese was the construction of the Great Wall. Many historians give the Emperor Chin Hwang-ti credit for the construction of the Wall, but portions of it were in existence before he came to the throne (221 B.C.). Repairs and extensions were made as late as the eighteenth century.

The part built by Hwang-ti had a two-fold purpose: (1) to act as a barrier against the barbarians, and (2) to serve as a monument to the union of China. He had just succeeded in breaking up the feudal system and had united China under a strong central government.

Hwang-ti's portion was built under the direction of an army engineer, General Meng Tien. His force consisted of 300,000 soldiers and hundreds of thousands of other laborers. The soldiers aided in the construction work when they were not needed to repel attacks. This portion of the Wall was completed about 204 B.C., after the death of Hwang-ti.

In later years Tartar activities caused the Wall to be extended, and new loops were built south of the original wall. In 423 A.D., a section about seven hundred miles in length was built running almost north and south along the western border of the province of Shensi. In 543 A.D., a second portion was built in Shensi, far south of the original wall. In 555 A.D., 1,800,000 men were used to build that part of the Wall near the present site of Peking. This is the wall usually seen by visitors to China.

Construction of Great Wall

The Great Wall, following all its curves, loops, and spurs, is about 2,550 miles long. The direct distance between its two ends is 1,145 miles. There were about 25,000 towers on the Wall and 15,000 detached watch-towers. The cross-section and the material used varied. At one point it was seventeen feet, six inches thick and sixteen feet high; it had two face walls of large brick, the space between being filled with earth and stones. On the Chinese side, the face wall was carried up three feet higher than the earth and stone filling, to form a parapet. On the Tartar side, the parapet was five feet high and cut down at frequent intervals with embrasures. The material used varied from solid masonry, with good stone and brickwork, down to a mere barrier of reeds and mud.

The Wall guards were organized into companies of one hundred and forty-five men under a commandant, who was responsible for a small group of watchtowers. A few mounted messengers were assigned to each compay. Several companies were grouped into "sections of the barrier" under a higher officer, who reported directly to the Governor of the province.

Training was not at all complete. Troops were disbanded when they were not needed; they were called to the colors when necessary. The troops who guarded the Great Wall might better be called settlers. They were given grants of land near the Wall and encouraged to marry. The result was that when a soldier's particular unit was not actually on guard on the Wall, he cultivated his land and had not time for training. At other times and places in the Empire, the low pay forced the soldier to seek additional employment, with the result that the Chinese army was composed of laborers and peasants who gave their spare moments, if they had any, to military exercises. The Emperor Taitsong's army of 900,000 soldiers had less value than 300,000 real soldiers.

At intervals along the Wall piles of dry reeds were kept ready for building signal fires to give warning of the approach of the Tartars. The strength of the attacking force was given by repetition of the fire signals. It was considered a serious offense for the soldiers at a watch tower to fail to transmit a signal received. The watch stations were about two and a half miles apart. rockets were also used for signalling.

There is no information that the Chinese had any general reserves back of the Wall, as the Romans had for their frontier defenses. The Chinese had defending troops at the gates; but between the gates the Wall served principally as an obstacle to small raiding parties, because the Tartars could not get their horses over it and small bands of barbarians without tools had difficulty in opening a passage through it. The Tartars were helpless without their horses and would not think of making a raid on foot.

The Wall did not prevent great invasions, as the Tartars in large bodies passed it many times. But such irruptions were made possible by the weakness or disloyalty of the defenders rather than by any weakness of the Wall itself. For example, when Genghis Khan (1162-1227) invaded China he bribed his way through a gate in the wall.

Military Transportation

In spite of the great number of boats on the canals and rivers, water transportation for military purposes seems to have been seldom used before the time of Kublai Khan (1216-94). About 280 A.D. a fleet on the Yang-tse-kiang was used in one of the civil wars. Again, in the year 969 A.D., an attempt was made to use boats in an expedition against the Cathayans, but low water so hindered the expedition that it failed.

The condition of the roads varied according to the character of the emperor. They were generally poor. Under the more energetic rulers, there were eras of road and bridge construction. Under poor rulers, and in periods of civil war, the roads were permitted to go to pieces. The Emperor Hwang-ti, who ordered the Great Wall built, constructed many roads. The Emperor Kaotsou (202-194 B.C.), founder of the Han Dynasty, used 100,000 men to construct a strategic road across the mountains to the city of Singafoo. Many suspension bridges were used on this road. The most notable was 150 yards long, wide enough for four horses to travel abreast, and 500 feet above the valley. It is said to be still in existence. The Emperor Chang-ti (76-89 A.D.)

spent 100,000 taels ($140,000) on the construction of a road to Cochin China, so that the tribute of that country could be sent by land to avoid pirates. Under the Emperor Tsin Wu-ti an engineer named Touyu, in 274 A.D., built a bridge across the great Hwang-Ho at Mongtsin. Later rulers permitted this bridge to go to pieces. Roads do not seem to have been paved. Military movements were usually slow on account of poor roads. This was especially so in the wars with the Tartars.

Early Rulers

For nearly a thousand years China was ruled by the Chou Dynasty (1122-256 B.C.). At that time the feudal system existed in the country. The subordinate princes warred with one another and paid merely nominal obedience to the emperor. The state of Tsin or Chin grew stronger and stronger until about 255 B.C., when the Prince of Tsin rebelled against the Chou Dynasty and overthrew it. He became emperor, and founded the Tsin Dynasty.

The great man of the Tsin Dynasty was Tsin Chin Hwang-ti (221-209 B.C.), who destroyed the feudal system, gave China a strong central government and started the Great Wall. His successor was overthrown in 206 B.C. by the first emperor of the Han Dynasty (206 B.C.-220 A.D.).

The early years of this dynasty were largely occupied by wars with the Tartars. A period of comparative peace and great prosperity followed.

The Emperor Wu-ti (140-86 B.C.) fought the Tartars almost continuously during the fifty-four years of his reign.

Advent of Tartars

"Tartars" was the general term applied to the barbarians who lived to the north and west of China. There were many distinct tribes. As these tribes, one after the other, gained the ascendancy among the Tartars, the name of that particular tribe was applied to the whole race. They were first called Hiung-nu (Huns), then Sien-Pi, and later Turks, Uigers, Cathayans, and Mongols. Their characteristics were very similar. They lived on horseback and moved from place to place with their flocks and herds in search of fresh pasture. Having no homes for their enemies to destroy and possessing greater mobility than the Chinese, who were mostly infantry, they retreated before the expeditions sent against them until the supplies of the Chinese were exhausted and they had to return to China.

The Tartars frequently raided Chinese territory in search of plunder. They preferred silk cloth and wine, but carried off almost anything that was easily portable. They were noted for their cruelty and frequently massacred the inhabitants of the Chinese cities. About the year 700 A.D., a Tartar leader named Mercho captured two cities in Shensi province and killed 90,000 inhabitants. In the year 947, the inhabitants of the city of Chang-teh Fu to the number of 100,000 were massacred by the Cathayans.

Wu-ti, sixth emperor of the Han Dynasty, came to the throne in 140 B.C., when he was sixteen years of age. The previous sixteen years had been peaceful and only minor raids by small bands of Tartars had disturbed the border. The Chinese had made good use of this interval of peace and the country was in a high state of material prosperity.

In 135 B.C. the Tartar king sent an envoy to ask for a Chinese princess in marriage and to express a desire for the continuance of the truce. In previous instances, such missions had often been the forerunner of war. Wu-ti yielded to the Tartar demands and peace continued for a short time.

War Against the Tartars

One of Wu-ti's advisors, Wang Ku, who had had personal experience on the border, advocated a policy of destroying the Tartars once for all rather than to remain exposed to their attacks. After much persuasion, Wu-ti decided to try this aggressive policy; and the man who had suggested the plan was placed in command of the Chinese expedition. An army of 300,000 men was raised and concentrated near the frontier.

As the Chinese were less mobile than the Tartars, Wang Ku tried to entice them into his reach by a strategem. A Chinese merchant was sent to offer the Tartars, for a consideration, the city of Ma Ch'eng, or "Horse City," one of the most contested places on the frontier. The Chinese army was placed in hiding so as to surround and destroy the Tartar king and his army when they came to take over the city.

The scheme almost succeeded. The Tartar king, with 100,000 of his best horsemen had passed the Great Wall and was within thirty miles of the city when he became suspicious. He went at once to the nearest tower on the Great Wall and captured its warden, who, to save his life, disclosed the whole scheme to the Tartars. The Tartars promptly retired from their exposed position and the scheme fell through. The warden was rewarded by a high position among the Tartars.

Wang Ku now tried to follow up and attack the Tartars; but they eluded him and the campaign ended without results. When Wu-ti heard of the failure of the scheme he ordered the arrest of Wang Ku, who, hearing of this order, followed the usual Chinese custom and committed suicide. Wu-ti's first effort against the Tartars was a complete failure.

The Tartars were inspired to fresh efforts; and their raids became more daring and almost incessant. In 127 B.C. they raided heavily the provinces of Kansuh and Shensi. To guard against this, Wu-ti resorted to a plan of forming military settlements in Shensi. He also improved the roads to that part of the country.

Wu-ti now appointed Wei Tsing to the command of a large army and sent him against the Tartars with instructions to engage them wherever they were found. Their successes had caused the Tartars to lose some of their habitual caution; and Wei Tsing succeeded in surrounding them. The Tartars, with the greater part of their forces, cut their way through the Chinese, but left behind their camp, baggage, women, children, and more than 15,000 warriors in the hands of the Chinese. A few months later, Wei Tsing again attacked the Tartars. He was victorious. These two battles caused the Chinese to recover confidence in themselves, which the previous Tartar successes had greatly impaired.

Offensive Against the Tartars

Encouraged by these victories, Wu-ti decided to carry the war into the enemy's territory. An army, consisting mostly of cavalry, was prepared and placed under Hokiuping, an experienced officer. As the Chinese had previously been mostly infantry, the Tartars were taken by surprise and offered little resistance. The expedition was successful Hokiuping traversed a large section of Tartar territory and returned to China with a great quantity of booty. His success illustrates the value of surprise in the use of a new arm.

A short time later, Hokiuping, with a larger force, again invaded Tartar territory and killed 30,000 Tartars. The Tartars felt that the Chinese successes were due to the incompetency of the Tartar leaders and became so threatening that two Tartar princes and their followers surrendered to the Chinese for saftey. They were disarmed and dispersed throughout several provinces.

In the Spring of 120 B.C., Wu-ti sent out a new expedition. This was divided into two columns, one under Wei Tsing and the other under Hokiuping. The Tartars retreated northward across the desert, and the Chinese followed. The Tartars sent their women and heavy baggage to a safe distance and turned to meet the Chinese. In the battle that followed, the Tartars were completely defeated and great numbers were killed. As a result of this campaign, no Tartars remained south of the desert; and half a million Chinese moved west to settle in the newly conquered territory. The year's fighting cost the Tartars 90,000 men. The Chinese lost 25,000 men and 100,000 horses.

The Tartar king now asked for peace, but refused to agree to all of China's demands. China was preparing a new expedition when her best general, Hokiuping, died. His loss was so great a blow that the proposed expedition was abandoned.

Wu-Ti Extends Kingdom

For the next few years Wu-ti was engaged in wars in other places. He conquered the provinces of Canton and Foochow, and a part of Korea. He was too busy to pay much attention to the Tartars. The Tartars were so exhausted that they were glad of the respite. Negotiations were carried on, but no satisfactory basis for peace was reached. About this time China lost her other great leader, Wei Tsing.

In 110 B.C. Wu-ti assembled an army of nearly 200,000 men near the border as though in preparation for an expedition against the Tartars; but he changed his mind and used it to reconquer the province of Liaotung, which had won its independence at the time the Tsin Dynasty fell. Wars with China's neighbors on the south and southwest again occupied Wu-ti's attention for a time; and when new expeditions were sent against the Tartars, they were unsuccessful. The advance guard of the first expedition was cut off and destroyed, which caused the Chinese army to retreat. The Tartars followed up this success by a raid through the province of Shensi, burning the towns and killing the inhabitants; but the death of the Tartar king caused the Tartar activity to cease for a time and gave Wu-ti a chance to prepare for the next phase of the struggle.

In 101 B.C., Wu-ti sent an army under Li Kwangli against the Tartars. This commander was so incompetent that his army was almost completely destroyed in the first battle, but the general escaped. To retrieve this disaster, a new expedition under Liling, grandson of Li Kwangli, was sent out. He invaded Tartar territory and was at first successful; but the Tartars won the third battle, and Liling started to retreat to China. He was soon surrounded by superior forces and surrendered. The Chinese consider it more honorable to take service under their conqueror than to return in disgrace to their own country. Liling accepted a high position in the Tartar army.

The next great Chinese expedition was sent out in 90 B.C. Li Kwangli was given another chance, and placed in command of this expedition. He was at first succesful and won several battles, but on his way back to China was surprised and defeated. He surrendered, and like his grandson, accepted service with the Tartars.

Value of the Offensive

This defeat ended Wu-ti's fifty years of war with the Tartars, as he died three years later without sending out any more expeditions. From this series of wars we should learn that victories should be followed up and the defeated enemy be given no chance to recover. In 120 B.C. the Chinese were completely victorious and another year's fighting probably would have meant the complete destruction of the Tartars, whereas thirty years later the struggle ended indecisively. These wars show the value of the offensive in war. China was unfortunate in the death of her great generals, Wei Tsing and Hokiuping.

The war against the Tartars of Liaotung was conducted by Taitsong and Chintsong, second and third emperors of the Sung Dynasty (906-1279 A.D.). The period just prior to the establishment of this dynasty was one of anarchy, and the empire was divided into several kingdoms. In his efforts to establish his authority over the whole empire, Taitsou, first emperor of this dynasty, had fought several wars, and had commenced one against the Prince of Han. A strong tribe of Tartars, allies of the Prince of Han, was now in possession of the province of Liaotung. After Taitsou's death, his successor, Taitsong, brought the war with the Prince of Han to a successful conclusion, and then decided to punish the Tartars for having been allied with Han.

Taitsong thought that he could conquer the king of Liaotung in a single campaign. Using his veterans of the war with Han, he invaded Liaotung, won some successes, and took several cities. Hearing that the Tartars were gathering their forces, he set out to defeat them in detail before their concentration was complete. The Tartar concentration was more advanced than Taitsong thought. While he was busy defeating one corps, which he had intercepted near the Kaoleang river on its march to the concentration area, he was attacked from the rear by the main body of the Tartars. The Chinese were completely defeated, with the loss of 10,000 men and all their baggage and equipment. The emperor escaped. This defeat shows how necessary it is to have complete information of the enemy.

Yeliu Hiuco, the Tartar general, was the outstanding figure of this war. At Kaoleang his division made the attack which completely routed the Chinese. Hiuco was distinguished by his tactical knowledge, his personal bravery, and other characteristics desirable in a leader. After Kaoleang, he was placed in command of a Tartar army. For the next twenty years, he was a thorn in the side of the Chinese.

Tartars Resume Offensive

There was little fighting for several years, but peace was not formally declared. Taitsong spent the interval in trying to plan the overthrow of Liaotung. In 985 A.D., he thought his opportunity had arrived. Korea asked for aid against the Tartars of Liaotung; and an offensive and defensive alliance against them was entered into by China and Korea.

The Chinese army promptly invaded Liaotung, took the Tartars by surprise and won several battles. The Chinese successes were checked by the arrival of Yeliu Hiuco. Near the fortress of Kikieou Koan, north of what is now Peking, Hiuco defeated the Chinese almost as badly as at Kaoleang. He promptly followed up his victory by pursuit and drove the Chinese army into the river Chaho. So many Chinese were drowned that "the corpses of the slain arrested the course of the river."

Other defeats for the Chinese followed. Finally, a Chinese army under General Panmei was de-

COMPARATIVE HISTORICAL CHART

Published by permission of W. W. Norton and Co.; from *The Making of Modern China*, by Owen and Eleanor Lattimore.

B.C.	WESTERN WORLD	DYNASTIES	CHINESE WORLD	B.C.
1800	Hammurabi BRONZE AGE	HSIA	NEOLITHIC AGE. Agricultural communities in Yellow River valley cultivated loess soil with stone tools. Domesticated dog and pig.	1800
1700			Hunting and fishing tribes in Yangtse valley.	1700
1600		SHANG		1600
1500	EGYPTIAN NEW EMPIRE Moses		BRONZE AGE. Primative Yellow River city states. Probable use of irrigation. Shang-inscribed bones give base line of history. Sheep goats domesticated.	1500
1400			Writing. Beautiful bronze castings. Potter's wheel. Stone carving. Silk culture and weaving. Wheeled vehicles.	1400
1300	Trojan War			1300
1200				1200
1100	IRON AGE			1100
1000	Solomon	CHOU	ANCIENT FEUDALISM. Expansion from Yellow River to Yangtse valley. "City and country" cells.	1000
900	Lycurgus		Increased irrigation. Eunuchs. Horse-drawn war chariots. 841 B.C. earliest authenticated date.	900
800	Carthage founded		Glass.	800
700	Hebrew prophets Greek lyric poets			700
600			IRON AGE Round coins. Magnetism known. CLASSICAL PERIOD. Confucius, Lao-tze.	600
500	Persian Wars Socrates		SUN TZU WU.	500
400	Plato Aristotle		Mencius.	400
300	Alexander Punic Wars		Bronze mirrors. BEGINNING OF EMPIRE. Great Wall.	300
200	Carthage and Corinth destroyed	CHIN	Palace architecture. Trade through Central Asia with Roman Empire. Ink.	200
100	Julius Caesar			100

A.D.				A.D.
	Birth of Christ Jerusalem destroyed	HAN	First Buddhist influences.	
100	Marcus Aurelius		Paper.	100
200		3 KINGDOMS	Tea.	200
300	Constantine Roman Empire divided	CHIN	Political disunity but cultural progress and spread.	300
400	Odoacer takes Rome	WEI / SUNG, CHI, LIANG, CHEN	Buddhism flourishing. Use of coal. Trade with Indo-China and Siam.	400
500	Justinian			500
600	Mohammed's Hegira	SUI	Large-scale unification. Grand Canal. ZENITH OF CULTURE. Chinese culture reaches	600
700	Moslems stopped at Tours	TANG	Japan. Turk and Tungus alliances. Revival of Confucianism weakens power of Buddhist	700
800	Charlemagne Alfred		monasteries. Mohammedanism. Cotton from India. Porcelain. First printed book.	800
900	Holy Roman Empire	5 DYNASTIES	State examinations organized. Rise of Khitan. Foot binding. Poetry, painting, sculpture.	900
1000	CRUSADES	LIAO / SUNG	Wang An-shih. Classical Renaissance. Paper money.	1000
1100		CHIN / SUNG	Rise of Jurchid. Compass. Navigation and mathematics.	1100
1200	Magna Carta		MONGOL AGE. Jenghis Khan. Marco Polo. Franciscans.	1200
1300	RENAISSANCE	YUAN	Operatic theater. Novels. Lamaism.	1300
1400	Printing in Europe Turks take Constantinople	MING	Yung Lo builds Peking. Period of restoration and stagnation.	1400
1500	AGE OF DISCOVERY		Portuguese traders arrive. Clash with Japan over Korea.	1500
1600	Religious Wars		Nurhachi	1600
1700		CHING	Critical scholarship. Canton open to Western trade.	1700
1800	American, French, Industrial Revolutions		Treaties with Western powers. Spread of Western culture. Taiping Rebellion.	1800
1900	First World War Russian Revolution Second World War	REPUBLIC	Boxer Rebellion. 1911 Revolution. Nationalist Revolution. Unification under Chiang Kai-shek. Japanese invasion and World War II.	1900

feated with a loss only slightly less than at Kikieou Koan. This left the Tartars masters of the situation and free to raid the border districts. Taitsong died and was succeeded by his son, Chintsong. Hiuco, the Tartar general, died soon after.

The Tartars resumed hostilities, but without Hiuco they were unsuccessful. They were repulsed by the governor of a border province. Chintsong led a large army to the border and the Tartars retreated. He was called away by a revolt in another part of the Empire and the invasion of Liaotung was postponed.

The Tartars resumed their raids into the border provinces. The Chinese were discouraged and feared to meet the Tartars in battle. Nevertheless, Chintsong collected a large army and in the year 1004 A.D. led it across the Hwang-Ho to attack the Tartars. The two armies came face to face, but the battle was never fought. Peace was concluded instead. The Tartars gave up a few captured towns and promised not to raid Chinese territory, in return for which the Chinese promised them an annual allowance of silk and money. Thus the Chinese obtained by purchase the peace which they had been unable to win by recourse to arms. Later, these Tartars conquered all of China, and founded the Manchu Dynasty. A real peace cannot be obtained by purchase.

Europeans Attack China

The foregoing campaigns comprise only operations between Chinese and Tartars or between

Chinese and other Chinese. Europeans did not encounter the Chinese until the Opium War, 1839-42. This was caused by the attempt of the Chinese to end the opium trade. Great Britain, to whom the traffic was a source of high profit, insisted upon being allowed to continue to sell opium to China. The British won. As a result, treaty ports were opened, Hong Kong was ceded to the English and a system of tariff oversight by the British was installed.

War between the Chinese and the British and the French, due to claims of various natures pressed by the foreigners and resisted by the Chinese, came in 1856-60. The European allies sent land and naval forces to take the forts at the mouth of the Peiho river, and to advance to Peking.

The Chinese were commanded by Sankolinsin. The Chinese force was much larger than the allied forces.

Anglo-French Offensive

On August 1, 1860, the allies landed unopposed near Peh-tang and entered that town early the next morning. On August 12th the allied forces moved against Sin-ho. They encountered some Chinese cavalry, and drove it off. The mud defenses of Sin-ho were then easily taken. The French pushed on at once and attacked Tang-ku, a fortified village southeast of Sin-ho; but they were repulsed. On the 14th, after bombarding

Tang-ku with thirty-six guns, the combined forces assaulted the city and captured it.

On August 21st, the allies attacked the forts at the mouth of the Peiho. The two upper forts were bombarded for three hours by forty-three field guns. An assault was then made on one of these forts north of the river. South China coolies carried the scaling ladders to help the British across the ditches and over the parapet. The Chinese resisted desperately, but the assault was successful. Of the garrison of 500 men, 400 were killed or captured.

Meanwhile, the fleet had been bombarding the two lower forts. A shell from the fleet caused an explosion in the magazine of the lower fort north of the river. Discouraged by this disaster and by the loss of the upper fort, the Chinese made little resistance; and the allies soon captured the fort and 2,000 Chinese soldiers. The forts on the south bank surrendered without further resistance. The day's fighting cost the British 22 killed and 179 wounded.

On August 23rd the march to Peking began. The Chinese caused some delay by opening negotiations for peace and then refusing to accept the allies' terms. Battles were fought at Chan-chiawan and at Pa-le-chiao Bridge. The Chinese numbered about 80,000 against 15,000 British and French; but the resistance of the Chinese was in vain.

The allies arrived before the walls of Peking on October 6th. After some negotiations, China yielded to the allied demands. A large indemnity was paid and the An-tiñg Gate and a part of the city walls adjacent to it were turned over to the allies. The British burned the emperor's summer palace near Peking, because the Chinese had tortured and mistreated British subjects whom they had captured. Thus a nation of 400,000,000 people was humiliated and its capital taken by about 15,000 men. The only parallel in history is the capture of Washington by a small British force in 1814.

A war of aggression and territory-grabbing, with Korea the main object, was begun against China by Japan in 1894. Japan won easily, but was prevented from enjoying the bulk of the fruits of her victory by England and Russia.

The Boxer Rebellion

During the so-called Boxer Rebellion in 1900, the Chinese, encouraged by the old Empress Tzu Hsi, in an effort to halt European penetration and influence, resisted an international army which fought its way from the coast to Peking. Foreigners, including members of the diplomatic corps, were besieged in the capital. The international army was composed of British, French, German, Russian, American and Japanese troops.

Then came the revolution of 1911, led by Sun Yat-Sen, which led to the overthrow of the Manchu dynasty, which had governed since the 17th century, and the establishment of the Chinese Republic. A military dictatorship and long continued civil war ensued. Finally, under General Chiang Kai-Chek the Kuomintang, or Nationalist, government was set up in Nanking in 1928.

The China "Incident"

Japan utilized distraught domestic conditions and grievances complained of on score of Nationalist resistance to Japanese territorial and trade ambitions to invade China in 1931. Manchuria was overrun and a puppet state set up there, headed by the former Manchu Emperor. Peking, renamed Peiping, was taken by the Japanese.

General war between the two China and Japan began in July, 1937. The Chinese valiantly resisted and, despite heavy handicaps and seizure of most of her important ports, continued to maintain effective opposition to complete Japanese victory, both before and after the commencement of World War II, which resulted in the Nationalist government becoming an ally of The United States, Great Britain, Russia and lesser associated states against the German-Japanese-Italian Axis.

This was the situation that prevailed in 1944, with General Chiang Kai-Shek's government occupying Chungking as the seat of the National Government.

The text of the *Art of War* has been transcribed from the translation by Lionel Giles, M.A., Assistant in the Department of Oriental Books and Manuscripts in the British Museum. It was published by Luzac & Co., London, in 1910. The critical notes of the translator, which comprise the larger portion of his book, have been summarized in the text whenever they had military value. Other translations in English and French lack both the accuracy and crystalline language which distinguish Dr. Giles. Grateful acknowledgment is made to Dr. Giles and Luzac & Co. for their generous permission to use his translation.

[ONE]

Laying Plans

SUN TZU said: The art of war is of vital importance to the state. It is a matter of life and death, a road either to safety or to ruin. Hence it is a subject of inquiry which can on no account be neglected.

The art of war is governed by five constant factors, to be taken into account in one's deliberations, when seeking to determine the conditions obtaining in the field.

These are: the Moral Law, Heaven, Earth, the Commander and Method and Discipline.

The Moral Law causes the people to be in complete accord with their ruler, so that they will follow him regardless of their lives, undismayed by any danger.

Heaven signifies night and day, cold and heat, times and seasons.

Earth comprises distances, great and small; danger and security; open ground and narrow passes; the chances of life and death.

The Commander stands for the virtues of wisdom, sincerity, benevolence, courage and strictness.

By *Method and Discipline* are to be understood the marshaling of the army in its proper subdivisions, the gradations of rank among the officers, the maintenance of roads by which supplies may reach the army, and the control of military expenditure.

These five heads should be familiar to every general. He who knows them will be victorious; he who knows them not will fail.

Therefore, in your deliberations, when seeking to determine the military conditions, let them be made the basis of a comparison, in this wise:

Seven Searching Questions

(1) Which of two sovereigns is imbued with the moral law?

(2) Which of two generals has most ability?

(3) With whom lie the advantages derived from heaven and earth?

(4) On which side is discipline most rigorously enforced?

(5) Which army is the stronger?

(6) On which side are officers and men most highly trained?

(7) In which army is there the greater constancy both in reward and punishment?

By means of these seven considerations I can forecast victory or defeat.

The general who harkens to my counsel and acts upon it, will conquer. Let such a one be retained in command! The general who harkens not to my counsel nor acts upon it, will suffer defeat. Let such a one be dismissed! While heeding the profit of my counsel, avail yourself also of any helpful circumstances over and beyond the ordinary rules. According as circumstances are favorable, one should modify one's plans.

[Sun Tzu is a practical soldier and wants no bookish theories. He cautions here not to pin one's faith on abstract principles. Tactics must be guided by the action of the enemy, as is well illustrated by Sir W. Fraser in his *Words on Wellington:* On the eve of the battle of Waterloo, Lord Uxbridge asked the Duke of Wellington what his plans were for the morrow, because, he explained, he might suddenly find himself Commander in Chief and would be unable to frame new plans in a critical moment. The Duke asked, "Who will attack first tomorrow—I or Bonaparte?" "Bonaparte," replied Lord Uxbridge. "Well," continued the Duke, "Bonaparte has not given me any idea of his projects; and as my plans will depend on his, how can you expect me to tell you what mine are?"]

Elemental Tactics

All warfare is based on deception. Hence, when able to attack, we must seem unable; when using our forces, we must seem inactive; when we are near, we must make the enemy believe that we are away; when far away, we must make him believe we are near. Hold out baits to entice the enemy. Feign disorder, and crush him.

If he is secure at all points, be prepared for him. If he is superior in strength, evade him. If your opponent is of choleric temper, seek to irritate him. Pretend to be weak, that he may grow arrogant.

If he is inactive, give him no rest. If his forces are united, separate them. Attack him where he is unprepared, appear where you are not expected. These military devices, leading to victory, must not be divulged beforehand.

The general who wins a battle makes many calculations in his temple ere the battle is fought. The general who loses a battle makes but few calculations beforehand. Thus do many calculations lead to victory, and few calculations to defeat: How much more do no calculation at all pave the way to defeat! It is by attention to this point that I can see who is likely to win or lose.

[TWO]

Waging War

SUN TZU said: In the operations of war, where there are in the field a thousand swift chariots, as many heavy chariots and a hundred thousand mail-clad soldiers, with provisions enough to carry them a thousand *li* [2.78 modern li make one mile] the expenditure at home and at the front, including entertainment of guests, small items such as glue and paint, and sums spent on chariots and armour, will reach the total of a thousand ounces of silver per day. Such is the cost of raising an army of 100,000 men.

> [It is interesting to note the similarity between early Chinese warfare and that of the Homeric Greeks. In each case, the war chariot was the nucleus around which was grouped the foot soldiers. Each light Chinese chariot was accompanied by 75 infantry, and each heavy chariot by 25 infantry, so that the whole army would be divided into a thousand battalions, each consisting of two chariots and 100 men.]

When you engage in actual fighting, if victory is long in coming, the men's weapons will grow dull and their ardour will be damped. If you lay siege to a town, you will exhaust your strength.

Again, if the campaign is protracted, the resources of the state will not be equal to the strain.

Now, when your weapons are dulled, your ardour damped, your strength exhausted and your treasure spent, other chieftains will spring up to take advantage of your extremity. Then no man, however wise, will rarely be able to avert the consequences that must ensue.

Thus, though we have heard of stupid haste in war, cleverness has never been associated with long delays. There is no instance of a country having been benefited from prolonged warfare.

Blitzkrieg

It is only one who is thoroughly acquainted with the evils of war who can thoroughly understand the profitable way of carrying it on. The skillful soldier does not raise a second levy, neither are his supply-wagons loaded more than twice. Bring war material with you from home, but forage on the enemy. Thus the army will have enough for its needs.

> [Once war is declared, the great general strikes immediately without waiting until every last detail is taken care of. This may seem audacious advice, but all great strategists, from Julius Caesar to Hitler, realized that time is of vital importance. "Too little, too late" was not one of their mottos.]

Poverty of the state exchequer causes an army

to be maintained by contributions from a distance. Contributing to maintain an army at a distance causes people to be impoverished.

On the other hand, the proximity of an army causes prices to go up; and high prices cause the people's substance to be drained away.

When their substance is drained away, the peasantry will be afflicted by heavy exactions.

Living at Enemy Expense

With this loss of subsistence and exhaustion of strength, the homes of the people will be stripped bare and three-tenths of their incomes will be dissipated; while government expenses for broken chariots, worn-out horses, breast-plates and helmets, bows and arrows, spears and shields, protective mantlets, draught-oxen and heavy wagons, will amount to four-tenths of its total revenue.

Hence a wise general makes a point of foraging on the enemy. One cartload of the enemy's provisions is equivalent to twenty of one's own, and likewise a single picul [about 133 pounds] of his provender is equivalent to twenty from one's own store.

In order to kill the enemy, men must be roused to anger; that there may be advantage from defeating the enemy, they must have their rewards.

Therefore in chariot fighting, when ten or more chariots have been taken, those should be rewarded who took the first. Our own flags should be substituted for those of the enemy, and the chariots mingled and used in conjunction with ours. The captured soldiers should be kindly treated and kept. This is called, using the conquered foe to augment one's own strength.

In war, then, let your great object be victory, not lengthy campaigns.

Thus it may be known that the leader of armies is the arbiter of the people's fate, the man on whom depends whether the nation shall be in peace or peril.

[THREE]

Attack by Stratagem

SUN TZU said: In the practical art of war, the best thing of all is to take the enemy's country whole and intact; to shatter and destroy it is not so profitable. So, too, it is better to capture an army entire than to destroy it, to capture a regiment, a detachment or a company entire than to annihilate them.

Hence to fight and conquer in all your battles is not supreme excellence; supreme excellence consists in breaking the enemy's resistance without fighting.

> [The elder Moltke's greatest triumph, the capitulation of the French at Sedan in 1870, was achieved practically without bloodshed. The Battle of France, May-June 1940, was the climax to a long succession of bloodless and practically bloodless victories for Hitler.]

Thus the highest form of generalship is to balk the enemy's plans. The next best is to prevent the junction of the enemy's forces. The next in order is to attack the enemy's army in the field. The worst policy of all is to besiege walled cities.

Siege Warfare

The rule is, not to besiege walled cities if it can possibly be avoided. The preparation of mantlets, movable shelters, and various implements of war, will take up three whole months; and the piling up of mounds [from which to attack] over against the walls will take three months more.

[Another sound piece of military theory. If the Boers had acted upon it in 1899 and refrained from dissipating their strength before Kimberley and Mafeking they would probably have been masters of the situation before the British were ready seriously to oppose them. The Germans beat their brains out before Stalingrad in 1943.]

The general who is unable to control his impatience will launch his men to the assault like swarming ants, with the result that one-third of his men are slain, while the town remains untaken. Such are liable to be the disastrous effects of a siege.

Therefore the skillful leader subdues the enemy's troops without any fighting; he captures their cities without laying siege to them; he overthrows their kingdom without lengthy operations in the field.

With his forces intact he will dispute the mastery of the empire, and thus, without losing a man, his triumph will be complete. This is the method of attacking by stratagem.

Advantage in Numbers

It is the rule in war, if our forces are ten to the enemy's one, to surround him; if five to one, to attack him; if twice as numerous, to divide our army into two.

If equally matched, we can offer battle; if slightly inferior in numbers, we can avoid the enemy; if quite unequal in every way, we can flee from him. Hence, though an obstinate fight may be made by a small force, in the end it must be captured by the larger force.

Now the general is the bulwark of the state: if the bulwark is complete at all points, the state will be strong; if the bulwark is defective, the state will be weak.

There are three ways in which a ruler can bring misfortune upon his army:

By commanding the army to advance or to retreat, being ignorant of the fact that it cannot obey. This amounts to hobbling the army.

By attempting to govern an army in the same way as he administers a kingdom, being ignorant of the conditions which obtain in an army. This causes restlessness among the soldiers.

By employing the officers of his army without discrimination, through ignorance of the military principle of adapting action to circumstances. This shakes the confidence of the soldiers.

But when the army is restless and distrustful, trouble is sure to come from other feudal princes. This is simply equivalent in results to bringing anarchy into the army and flinging victory away.

Thus we may know that there are five essentials for victory:

He will win who knows when to fight and when not to fight.

He will win who knows how to handle both superior and inferior forces.

He will win whose army is animated by the same spirit throughout all ranks.

He will win who, prepared himself, waits to take the enemy unprepared.

He will win who has military capacity and is not interfered with by the sovereign.

Victory lies in the knowledge of those five points.

Hence the saying: If you know the enemy and know yourself, you need not fear the result of a hundred battles. If you know yourself, but not the enemy, for every victory gained you will also suffer a defeat. If you know neither the enemy nor yourself, you will succumb in every battle.

[FOUR]

Tactical Dispositions

SUN TZU said: The good fighters of old first put themselves beyond the possibility of defeat and then waited for an opportunity of defeating the enemy.

To secure ourselves against defeat lies in our own hands, but the opportunity of defeating the enemy is provided by the enemy himself.

Thus the good fighter is able to secure himself against defeat, but cannot make certain of defeating the enemy.

Hence the saying: One may *know* how to conquer without being able to do it.

Security against defeat implies defensive tactics; ability to defeat the enemy means taking the offensive.

Standing on the defensive indicates insufficient strength; attacking, a superabundance of strength.

The general who is skilled in defense, in effect, hides in the most secret recesses of the earth; he who is skilled in attack flashes forth from the topmost heights of heaven. Thus on the one hand we have ability to protect ourselves; on the other, a victory that is complete.

To see victory only when it is within the ken of the common herd is not the acme of excellence. Neither is it the acme of excellence if you conquer and the whole empire says, "Well done!"

To lift an autumn leaf is no sign of great strength; to see sun and moon is no sign of sharp sight; to hear the noise of thunder is no sign of a quick ear. What the ancients called a clever fighter is one who not only wins, but excels in winning with ease.

Hence his victories bring him neither reputation for wisdom nor credit for courage. He wins his battles by making no mistakes. Avoidance of mistakes establishes the certainty of victory, for it means conquering an enemy that is already defeated.

To Avoid Defeat

Hence the skillful fighter puts himself into a position which makes defeat impossible, and does not miss the moment for defeating the enemy.

Thus it is that in war the victorious strategist seeks battle after his plans indicate that victory is possible under them, whereas he who is destined to defeat first fights without skillful planning and expects victory to come without planning.

The consummate leader cultivates the moral law, and strictly adheres to method and discipline. Thus it is in his power to control success.

In respect of military method, we have: First, measurement; second, estimation of quantity; third, calculation; fourth, balancing of chances; fifth, victory.

Measurement owes its existence to earth; estimation of quantity to measurement; calculation to estimation of quantity; balancing of chances to calculation; and victory to balancing of chances.

> [The first of these terms would seem to be terrain appreciation, from which an estimate of the enemy's strength can be formed, and calculations of relative strength made on the data thus obtained; we are thus led to a general comparison of the enemy's chances with our own; if the latter turn the scale, then victory ensues.]

A victorious army opposed to a routed one, is as a pound's weight placed in the scale against a single grain. The onrush of a conquering force is like the bursting of pent-up waters into a chasm a thousand fathoms deep. So much for tactical dispositions.

[FIVE]

Use of Energy

SUN TZU said: The control of a large force is the same in principle as the control of a few men. It is merely a question of dividing up their numbers.

Fighting with a large army under your command is nowise different from fighting with a small one. It is merely a question of instituting signs and signals.

To ensure that your whole host may withstand the brunt of the enemy's attack and remain unshaken is effected by direct and indirect maneuvers.

That the impact of your army may be like a grindstone dashed against an egg. That is effected by the science involving contacts between weak points and strong.

In all fighting, the direct method may be used for joining battle, but indirect methods will be needed in order to insure victory.

Indirect tactics, efficiently applied, are inexhaustible as heaven and earth, unending as the flow of rivers and streams; like the sun and moon, they end their course but to begin anew; like the four seasons, they pass to return once more.

There are not more than five musical notes, [probably meaning, only five employed in Chinese music of the era] yet the combinations of these five give rise to more melodies than probably can ever be heard. There are not more than three primary colors, yet in combination they produce more hues than can ever be seen. There are not more than five tastes, [sour, acrid, salt, sweet, bitter] yet combinations of them yield more flavors than can ever be tasted.

In battle, there are not more than two methods of attack—the direct and indirect; yet these two in combination give rise to an endless series of maneuvers. The direct and indirect lead on to each other in turn. It is like moving in a circle—you never come to an end. Who can exhaust the possibilities of their combination?

The onset of troops is like the rush of a torrent which will even roll stones along its course.

The quality of decision is like the well-timed swoop of a falcon which enables it to strike and destroy its victim.

Therefore the good fighter will be terrible in his onset, and prompt in his decision.

Decision Releases Force

Energy may be likened to the bending of a cross-bow; decision, to the releasing of the trigger.

Amid the turmoil and tumult of battle, there may be seeming disorder and yet no real disorder at all. Amid confusion and chaos, your array may be without apparent head or tail, yet it will be proof against defeat.

Simulated disorder postulates perfect discipline;

simulated fear postulates courage; simulated weakness postulates strength.

Hiding order beneath the cloak of disorder is simply a question of subdivision; concealing courage under a show of timidity presupposes a fund of latent energy; masking strength with weakness is to be effected by tactical dispositions.

Thus one who is skillful at keeping the enemy on the move maintains deceitful appearances, according to which the enemy will act.

By holding out baits, he keeps him on the march; then with a body of picked men he lies in wait for him.

The clever combatant looks to the effect of combined energy, and does not require too much from individuals. Hence his ability to pick out the right men and to utilize combined energy.

When he utilizes combined energy, his fighting men become as it were like unto rolling logs or stones. For it is the nature of a log or stone to remain motionless on level ground, and to move when on a slope; if four cornered, to come to a standstill, but if round-shaped to go rolling down.

Thus the energy developed by good fighting men is as the momentum of a round stone rolled down a mountain thousands of feet in height. So much on the subject of energy.

[SIX]

Weak Points and Strong

SUN TZU said: Whoever is first in the field and awaits the coming of the enemy, will be fresh for the fight; whoever is second in the field and has to hasten to the battle, will arrive exhausted.

Therefore the clever combatant imposes his will on the enemy, but does not allow the enemy's will to be imposed on him.

By holding out advantages to him he can cause the enemy to approach of his own accord; or by inflicting damage he can make it impossible for the enemy to draw near.

If the enemy is taking his ease he can harass him; if well supplied he can starve him out; if quietly encamped, he can force him to move.

Appear at points which the enemy must hasten to defend; march swiftly to places where you are not expected.

An army may march great distances without distress if it marches through country where the enemy is not.

You can be sure of succeeding in your attacks if you attack places which are not defended. You can insure the safety of your defense if you hold only positions that cannot be attacked.

Hence the general is skillful in attack whose opponent does not know what to defend; and he is skillful in defense whose opponent does not know what to attack. [An aphorism which puts the whole art of war in a nutshell.]

Subtlety and Secrecy

O divine art of subtlety and secrecy! Through you we learn to be invisible, through you inaudible; and hence hold the enemy's fate in our hands.

You may advance and be absolutely irresistible if you make for the enemy's weak points; you may retire and be safe from pursuit if your movements are more rapid than those of the enemy.

If we wish to fight the enemy can be forced to an engagement even though he be sheltered behind a high rampart and a deep ditch. All we need to do is to attack some other place which he will be obliged to relieve.

If we do not wish to fight, we can prevent the enemy from engaging us even though the lines of our encampment be merely traced on the ground. All we need to do is to throw something unused and unaccountable in his way.

By discovering the enemy's dispositions and remaining invisible ourselves, we can keep our forces concentrated while the enemy must be divided.

We can form a single united body, while the enemy must split up into fractions. Hence there will be a whole pitted against separate parts of a whole, which means that we shall be many in collected mass to the enemy's separate few, amongst his separated parts.

And if we are thus able to attack an inferior force with a superior one, our opponents will be in dire straits.

Battle-site Secrecy

The spot where we intend to fight must not be made known; for then the enemy will have to prepare against a possible attack at several different points; and his forces being thus distributed in many directions, the numbers we shall have to face at any given point will be proportionately few.

> [Sheridan once explained the reason of Grant's victories by saying that "while his opponents were kept busy wondering what he was going to do, *he* was thinking most of what he himself was going to do."]

For should the enemy strengthen his van, he will weaken his rear; should he strengthen his rear, he will weaken his van; should he strengthen his left, he will weaken his right; should he strengthen his right, he will weaken his left. If he sends reinforcements everywhere, he will be everywhere weak.

> [Frederick the Great, in his *Instructions to his Generals**, says "Those generals who have had but little experience attempt to protect every point; while those who are better ac-

* The Military Service Publishing Company.

quainted with their profession, guard against decisive blows at decisive points, and acquiesce in smaller misfortunes to avoid greater." In other words, keep away from sideshows.]

Numerical weakness comes from having to prepare against possible attacks; numerical strength, from compelling our adversary to make these preparations against us.

Knowing the place and time of the coming battle, we may concentrate from great distances in order to fight.

But if neither time nor place be known, then the left wing will be impotent to succor the right, the right equally impotent to succor the left, the van unable to relieve the rear, or the rear to support the van. How much more so if the furthest portions of the army are anything under a hundred *li* apart, and even the nearest are separated by several *li*.

Though according to my estimate the soldiers of Yüeh exceed our own in number, that shall advantage them nothing in the matter of victory. I say then that victory can be achieved.

Though the enemy be stronger in numbers, we may prevent him from fighting. Scheme so as to discover his plans and the likelihood of their success.

Seeking Enemy's Weakness

Rouse him, and learn the principle of his activity or inactivity. Force him to reveal himself, so as to find out his vulnerable spots.

Carefully compare the opposing army with our own, so that you may know where strength is superabundant and where it is deficient.

In making tactical dispositions, the highest pitch you can attain is to conceal them; conceal your dispositions and you will be safe from the prying of the subtlest of spies, from the machinations of the wisest brains.

How victory may be produced by this from the enemy's own tactics is what the multitude cannot comprehend.

All men can see these tactics whereby I conquer, but what none can see is the strategy out of which victory is evolved.

Do not repeat the tactics which have gained you one victory, but let your methods be regulated by the infinite variety of circumstances.

Tactics Are Fluid

Military tactics are like unto water, for water in its natural course runs away from high places and hastens downwards. So in war, the way to avoid what is strong is to strike what is weak.

Water shapes its course according to the ground over which it flows; the soldier works out his victory in relation to the foe whom he is facing.

Therefore, just as water retains no constant shape, so in warfare there are no constant conditions.

He who can modify his tactics in relation to

his opponent and thereby succeed in winning, may be called a heaven-born captain.

The five elements [water, fire, wood, metal, earth] are not always equally prominent; the four seasons make way for each other in turn. There are short days and long; the moon has its periods of waning and waxing.

[SEVEN]

Maneuvering An Army

SUN TZU said: In war, the general receives his commands from the sovereign.

Having collected an army and concentrated his forces, he must blend and harmonize the different elements thereof before pitching his camp.

After that, comes tactical maneuvering, than which there is nothing more difficult. The difficulty of tactical maneuvering consists in turning the devious into the direct and misfortune into gain.

> [Signal examples of this saying are afforded by the two famous passages across the Alps—that of Hannibal, which laid Italy at his mercy, and that of Napoleon two thousand years later, which resulted in the great victory of Marengo. The German's use of the "impenetrable" Ardennes in 1940 turned the devious into the direct with a vengeance.]

Thus, to take a long circuitous route, after enticing the enemy out of the way, and though starting after him to contrive to reach the goal before him, shows knowledge of the artifice of *deviation.*

Maneuvering with an army is advantageous; with an undisciplined multitude, most dangerous.

If you set a fully equipped army in march in

order to snatch an advantage, the chances are that you will be too late. On the other hand, to detach a flying column for the purpose involves the sacrifice of its baggage and stores.

Thus, if you order your men to make forced marches without halting day or night, covering double the usual distance at a stretch, and doing a hundred *li* in order to wrest an advantage, the leaders of your three divisions will fall into the hands of the enemy.

The stronger men will be in front, the jaded ones will fall behind, and by this only one-tenth of your army will reach its destination.

If you march fifty *li* in order to outmaneuver the enemy, you will lose the leader of your first division, and only half your force will reach its goal.

If you march thirty *li* with the same object, two-thirds of your army will arrive.

Losing Tactics

We may take it then that an army without its baggage train is lost; without provisions it is lost; without bases of supply it is lost.

We cannot enter into alliances until we are acquainted with the designs of our neighbors.

We are not fit to lead an army on the march unless we are familiar with the face of the country—its mountains and forests, its pitfalls.

We shall be unable to turn natural advantages to account unless we make use of local guides.

In war, practice dissimulation, and you will succeed. Move only if there is a real advantage to be gained.

Whether to concentrate or to divide your troops must be decided by circumstances.

Let your rapidity be that of the wind, your compactness that of the forest. In raiding and plundering be like fire, in immovability like a mountain.

Keep your plans dark and impenetrable as night and when you move, fall like a thunderbolt.

When you plunder a countryside, let the spoil be divided amongst your men; when you capture new territory, cut it up into allotments for the benefit of the soldiery.

Ponder and deliberate before you make a move.

He will conquer who has learnt the artifice of deviation. Such is the art of maneuvering.

Gongs, Drums, Banners, Flags

The *Book of Army Management* [an unknown book] says: On the field of battle the spoken word does not carry far enough: hence the institution of gongs and drums. Nor can ordinary objects be seen clearly enough; hence the institution of banners and flags.

Gongs and drums, banners and flags are means whereby the ears and eyes of the host may be focussed on one particular point.

The host thus forming a single united body, it is impossible either for the brave to advance alone, or for the cowardly to retreat alone. This is the art of handling large masses of men.

In night-fighting, then, make much use of signal fires and drums, and in fighting by day of flags and banners as a means of guiding your men through their ears and eyes.

A whole army may be robbed of its spirit; a commander-in-chief may be robbed of his presence of mind.

Now a soldier's spirit is keenest in the morning; by noonday it has begun to flag and in the evening his mind is bent only on returning to camp.

A clever general, therefore, avoids an army when its spirit is keen, but attacks it when it is sluggish and inclined to retreat. This is the art of studyng moods.

Self Possession and Strength

Disciplined and calm, to await the appearance of disorder and hubbub amongst the enemy—this is the art of retaining self possession.

To be near the goal while the enemy is still far from it, to wait at ease while the enemy is toiling and struggling, to be well fed while the enemy is famished—this is the art of husbanding one's strength.

To refrain from intercepting an enemy whose ranks are in perfect order, to refrain from attacking an army drawn up in calm and confident array—this is the art of studying circumstances.

It is a military axiom not to advance uphill

against the enemy, nor to oppose him when he comes downhill.

Do not pursue an enemy who simulates flight; do not attack soldiers whose temper is keen.

Do not swallow a bait offered by the enemy. Do not interfere with an army that is retreating into its own territory.

When you surround an army leave an outlet free. Do not press a desperate foe too hard.

Such is the art of warfare.

⟦ EIGHT ⟧

Variation of Tactics

SUN TZU said: In war, the general receives his commands from the sovereign, collects his army and concentrates his forces.

When in difficult country do not encamp. In country where high roads intersect join hands with your allies. Do not linger in dangerously isolated positions. In hemmed-in situations where you are you must resort to stratagem. In a desperate position, you must fight.

There are roads which must not be followed, armies which must not be attacked, towns which must not be besieged, positions which must not be contested, commands of the sovereign which must not be obeyed.

Flexible Tactics

The general who thoroughly understands the advantages that accompany variation of tactics knows how to handle his troops.

The general who does not understand these may be well acquainted with the configuration of the country, yet he will not be able to turn his knowledge to practical account.

The student of war who is unversed in the art of varying his plans, even though he be acquainted with the Five Advantages will fail to make the best use of his men.

Hence in the wise leader's plans, considerations of advantage will be blended. If our expectation of advantage be tempered in this way, we may succeed in accomplishing the essential part of our schemes.

If, on the other hand, in the midst of difficulties we are always ready to seize an advantage, we may extricate ourselves from misfortune.

Reduce hostile chiefs by inflicting damage on them; make trouble for them, and keep them constantly engaged; hold out specious allurements, and make them rush to any given point.

Preparedness Pays

The art of war teaches us to rely not on the likelihood of the enemy not coming, but on our own readiness to receive him; not on the chance of his not attacking, but rather on the fact that we have made our position unassailable.

There are five dangerous faults which may affect a general: Recklessness, which leads to destruction; cowardice, which leads to capture; a hasty temper that can be provoked by insults; a delicacy of honor that is sensitive to shame; over-solicitude for his men, which exposes him to worry and trouble. These are the five besetting sins of a general, ruinous to the conduct of war.

When an army is overthrown and its leader slain, the cause will surely be found among the five dangerous faults. Let them be a subject of meditation.

[NINE]

THE ARMY ON THE MARCH

SUN TZU said: We now come to the question of encamping the army and observing signs of the enemy. Pass quickly over mountains [which are barren of fodder] and keep in the neighborhood of valleys.

Camp in high places. Do not climb heights in order to fight. So much for mountain warfare.

After crossing a river, you should get far away from it [in order to tempt the enemy to follow you across].

When an invading force crosses a river in its onward march, do not advance to meet it in mid-stream. It will be best to let the army get across and then deliver your attack.

If you are anxious to fight, you should not go to meet the invader near a river which he has to cross [for fear of preventing his crossing].

Moor your craft higher up than the enemy and facing the sun. Do not move upstream to meet the enemy. So much for river warfare.

Advantages of Terrain

In crossing marshes, your sole concern should be to get over them quickly, without any delay.

If forced to fight in a marsh, you should have the water and grass near you, and get your back to a clump of trees. So much for marshes.

In dry, level country, take up an easily accessible position with rising ground to your right and on your rear, so that the danger may be in front, and safety lie behind. So much for flat country.

These are the four useful branches of military knowledge which enabled the Yellow Emperor [possibly one of the Chin dynasty, 256-207 B.C., from which was derived the name "China"] to vanquish four several sovereigns.

In battle and maneuvering all armies should prefer high ground to low and sunny places to dark. If you are careful of your men, and camp on hard ground, the army will be free from diseases, and this will spell victory.

When you come to a hill or a bank, occupy the sunny side, with the slope on your right rear. Thus you will at once act for the benefit of your soldiers and utilize the natural advantages of the ground.

When, in consequence of heavy rains up-country, a river which you wish to ford is swollen and flecked with foam, you must wait until it subsides. Country in which there are precipitous cliffs with torrents running between, deep natural hollows, confined places, tangled thickets, quagmires and crevasses, should be left with all possible speed and not even approached.

While we keep away from such places, we should try to get the enemy to approach them; while we face them, we should let the enemy have them on his rear.

Ambushes and Spies

If in the neighborhood of your camp there should be hilly country, ponds surrounded by aquatic grass, hollow basins filled with reeds, or woods with thick undergrowth, they must be carefully searched; for these are places where men in ambush or spies are likely to be lurking.

When the enemy is close at hand and remains quiet, he is relying on the natural strength of his position.

When he keeps aloof and tries to provoke a battle, he is anxious for the other side to advance.

If his place of encampment is easy of access, he is tendering a bait.

Movement amongst the trees of a forest shows that the enemy is advancing. The appearance of a number of screens [to fake an ambush] in the midst of thick grass means that the enemy wants to make us suspicious.

The rising of birds in their flight is the sign of an ambuscade. Startled beasts indicate that a sudden attack is coming.

When there is dust rising in a high column, it is the sign of chariots advancing; when the dust is low, but spread over a wide area, it betokens the approach of infantry. When it branches out in different directions, it shows that parties have been sent out to collect firewood. A few clouds of dust moving to and fro signify that the army is camping.

Placatory words and increased preparations are signs that the enemy is about to advance. Violent language and driving forward as if to the attack may be signs that he will retreat.

Military Intelligence

When light chariots come out and take up a position on the wings, it is a sign that the enemy is forming for battle.

Peace proposals unaccompanied by a sworn covenant indicate a plot.

When there is much running about it means that the critical moment has come.

When some of the enemy are seen advancing and some retreating, it is a lure.

When soldiers stand leaning on their spears, they are faint from want of food.

If those who are sent to draw water begin by drinking themselves, the army is suffering from thirst.

If the enemy sees an advantage to be gained and makes no effort to secure it, the soldiers are exhausted.

If birds gather on any spot, it is unoccupied. Clamour by night betokens nervousness.

Signs of Significance

If there is disturbance in the camp [as indicated by observation, reports of spies or otherwise], the general's authority is weak. If the banners and flags are shifted about, sedition is afoot. If the officers are angry, the men are weary.

When an army feeds its horses with grain and kills its cattle for food, and when the men do not hang their cooking pots over the camp-fires, showing that they will not return to their tents, you may know that they are determined to fight to the death.

The sight of men whispering together in small knots and speaking in subdued tones points to dissatisfaction amongst the rank and file.

Too frequent rewards signify that the enemy is at the end of his resources [because, when an army is hard pressed there is always fear of mutiny and lavish rewards are given to keep the men happy]. Too many punishments betray a condition of dire distress [because in such conditions discipline is relaxed and unwonted severity is necessary to keep the men to their duty].

To begin by bluster, but afterwards to take fright at the enemy's numbers, shows supreme lack of intelligence.

When envoys are sent with compliments in their mouths, it is a sign that the enemy wishes for a truce.

If the enemy's troops march up angrily and remain facing yours for a long time without either joining battle or taking themselves off again, the situation is one that demands great vigilance and circumspection.

If our troops are no more in number than the enemy, that is amply sufficient; it means that no direct attack may be made. What we can do is simply to concentrate all our available strength, keep a close watch on the enemy, and obtain reinforcements.

He who exercises no forethought, but makes light of his opponents risks being captured by them.

If soldiers are punished before they have grown attached to you, they will not prove submissive; and unless submissive, they will be practically useless. If, when the soldiers have become attached to you, punishments are not enforced, they will still be useless.

Therefore soldiers must be treated in the first instance with humanity, but kept under control by iron discipline. This is one of the certain roads to victory.

If in training soldiers commands are habitually enforced, the army will be well disciplined.

If a general shows confidence in his men but always insists on his orders being obeyed, the gain will be mutual.

[TEN]

Classification of Terrain

SUN TZU said: We may distinguish six kinds of terrain: Accessible ground, entangling ground, temporizing ground, narrow passes, precipitous heights and positions at a great distance from the enemy.

Accessible. Ground which can be freely traversed by both sides. With regard to ground of this nature, be before the enemy in occupying the raised and sunny spots, and carefully guard your line of supplies. Then you will be able to fight with advantage.

Entangling. Ground which can be abandoned but is hard to reoccupy. From a position of this sort, if the enemy is unprepared, you may sally forth and defeat him. But if the enemy is prepared for your coming, and you fail to defeat him, then, return being impossible, disaster will ensue.

Temporizing. Ground whereon the position is such that neither side will gain by making the first move. In a position of this sort, even though the enemy should offer us an attractive bait, it will be advisable not to stir forth, but rather to retreat, thus enticing the enemy in his turn; then, when part of his army has come out, we may deliver our attack with advantage.

Narrow passes. If you can occupy them first,

let them be strongly garrisoned and await the advent of the enemy. Should the enemy forestall you in occupying a pass, do not go after him if the pass is fully garrisoned, but only if it is weakly garrisoned.

Heights. You should occupy raised and sunny spots, and there wait for him to come up. If the enemy has occupied them before you, do not follow him, but retreat to entice him away.

If you are situated at a great distance from the enemy, and the strength of the two armies is equal, it is not easy to provoke a battle, and fighting will be to your disadvantage.

These six are the principles of terrain. The general who holds a responsible post must study them. [Terrain is treated more fully in the next chapter.]

Calamities Due to Error

An army is exposed to six severe calamities not arising from natural causes, but from faults for which the general is responsible. These are: flight, insubordination, collapse, ruin, disorganization and rout.

Other conditions being equal, if one force is hurled against another ten times its size, the result will be the *flight* of the former.

When the common soldiers are too strong and their officers too weak, the result is *insubordination.* When the officers are too strong and the common soldiers too weak, the result is *collapse.*

When the higher officers are angry and insubordinate, and on meeting the enemy give battle independently, on their own account from a feeling of resentment, before the commander-in-chief can tell whether or not he is in a position to fight, the result is *ruin.*

When the general is weak and without authority; when his orders are not clear and distinct; when there are no fixed duties assigned to officers and men, and the ranks are formed in a slovenly haphazard manner, the result is utter *disorganization.*

When a general, unable to estimate the enemy's strength, allows an inferior force to engage a larger one, or hurls a weak detachment against a powerful one, and neglects to place picked soldiers in the front rank, the result must be a *rout.*

These are six ways of courting defeat, which must be carefully noted by the general in active command and service.

Tests of Good Generalship

The natural formation of the country is the soldier's best ally; but a power of estimating the adversary, of controlling the forces of victory, and of shrewdly calculating difficulties, dangers and distances, constitutes the test of a great general.

He who knows these things, and in fighting puts his knowledge into practice, will win his battles. He who knows them not, will surely be defeated.

If fighting is reasonably sure to result in victory, then you must fight, even though the ruler forbid it; if fighting promises not to result in victory, then you must not fight, even at the ruler's bidding.

The general who advances without coveting fame and retreats without fearing disgrace, whose only thought is to protect his country and do good service for his sovereign, is the jewel of the kingdom.

Regard your soldiers as your children, and they will follow you wherever you may lead. Look on them as your own beloved sons, and they will stand by you even unto death.

If, however, you are indulgent, but unable to make your authority felt; kind-hearted, but unable to enforce your commands; and incapable, moreover, of quelling disorder, then your soldiers must be likened to spoiled children. They are useless for any practical purpose.

If we know that our own men are in a condition to attack, but are unaware that the enemy is not open to attack, we have gone only halfway towards victory.

If we know that the enemy is open to attack, but are unaware that our own men are not in a condition to attack, we have gone only halfway towards victory.

If we know that the enemy is open to attack, and also know that our own men are in a condition to attack, but are unaware that the nature of the ground makes fighting impracticable, we have gone only halfway towards victory.

Hence the experienced soldier, once in motion, is never bewildered. Once he has broken camp, he is never at a loss.

Hence the saying: If you know the enemy and know yourself, your victory will not stand in doubt; if you know Heaven and know Earth, you may make your victory complete.

[ELEVEN]

The Nine Situations

SUN TZU said: The art of war recognizes nine varieties of ground: Dispersive ground, facile ground, contentious ground, open ground, ground of intersecting highways, serious ground, difficult ground, hemmed-in ground and desperate ground.

Dispersive. Ground whereon a chieftain is fighting in his own territory.

Facile. Ground whereon he has penetrated into hostile territory, but to no great distance, it is facile ground.

Contentious. Ground, the possession of which imports great advantage to either side.

Open. Ground on which each side has liberty of movement.

Intersecting highways. Ground which forms the key to three contiguous states, so that he who occupies it first has most of the Empire at his command.

Serious. Ground whereon an army has penetrated into the heart of a hostile country, leaving a number of fortified cities in his rear.

Difficult. Ground that includes mountain forests, rugged steeps, marshes and fens—all country that is hard to traverse.

Hemmed-in. Ground which is reached through narrow gorges and from which we can retire only by tortuous paths, so that a small number of the enemy would suffice to crush a large body of our men.

Desperate. Ground on which we can only be saved from destruction by fighting without delay.

Using Terrain

On dispersive ground, therefore, fight not.

On facile ground, halt not.

On contentious ground, attack not.

On open ground, do not try to block the enemy's way.

On ground of intersecting highways, join hands with your allies.

On serious ground, gather in plunder.

On desperate ground, fight.

On hemmed-in ground, resort to stratagem.

[This is exactly what Hannibal did when he was hemmed in among the mountains on the road to Casilinum and to all appearances trapped by Fabius. According to Polybius and Livy, when night came on he fastened faggots to the horns of some 2000 oxen, set them afire and drove the terrified animals toward the passes held by the enemy. The Romans were so alarmed by the moving lights that they withdrew and Hannibal's army passed safely through the defile. The remarkable thing about this strategem is that one almost identical was used by T'ien Tan in China in 279 B.C., exactly 62 years before Hannibal used it.]

Those who of old were called skillful leaders knew how to drive a wedge between the enemy's front and rear; to prevent co-operation between his large and small divisions; to hinder the good troops from rescuing the bad, the officers from rallying their men.

Tactics of Skillful Leaders

When the enemy's men were scattered, they prevented them from concentrating; even when their forces were united, they managed to keep them in disorder.

When it was to their advantage, they made a forward move; when otherwise, they stopped still.

When asked how to cope with a great host of the enemy in orderly array and on the point of marching to the attack, I should say: "Begin by seizing something which your opponent holds dear; then he will be amenable to your will."

Rapidity is the essence of war; take advantage of the enemy's unreadiness, make your way by unexpected routes, and attack unguarded spots.

The following are principles to be observed by an invading force: the further you penetrate into a country, the greater will be the solidarity of your troops, and thus the defenders will not prevail against you.

Make forays in fertile country in order to supply your army with food.

Carefully study the well-being of your men, and do not overtax them. Concentrate your energy and hoard your strength. Keep your army continually on the move and devise unfathomable plans.

Throw your soldiers into positions whence there is no escape, and they will prefer death to flight. Officers and men alike will put forth their uttermost strength.

When Troops Fight Hard

Soldiers when in desperate straits lose the sense of fear. If there is no place of refuge, they will stand firm. If they are in the heart of a hostile country, they will show a stubborn front. If there is no help for it, they will fight hard.

Thus, without waiting to be marshaled, the soldiers will be constantly on the *qui vive;* without waiting to be asked, they will do your will; without restrictions, they will be faithful; without giving orders, they can be trusted.

Prohibit seeking for omens, and do away with superstitious doubts. Then, until death comes, no apparently predestined calamity need be feared.

If our soldiers are not overburdened with money, it is not because they have a distaste for riches; if their lives are not unduly long, it is not because they are disinclined to longevity.

On the day they are ordered out to battle, your soldiers may weep, those sitting up bedewing their garments, and those lying down letting the tears run down their cheeks. But let them once be brought to bay, and they will display the courage of any of our heroes.

Shuai-jan Tactics

The skillful tactician may be likened to the *shuai-jan*, a snake that is found in the Ch'ang mountains. Strike at its head and you will be attacked by its tail; strike at its tail, and you will be attacked by its head; strike at its middle, and you will be attacked by head and tail both.

Asked if an army can be made to imitate the *shuai-jan*, I should answer, yes. For the men of Wu and the men of Yüeh are enemies; yet if they are crossing a river in the same boat and are caught by a storm, they will come to each other's assistance just as the left hand helps the right.

Hence it is not enough to put one's trust in the tethering of horses, and the burying of chariot wheels in the ground.

> [These quaint devices to prevent one's army from running away recall the Athenian hero Sophanes, who carried an anchor with him at the battle of Plataea, 429 B.C., and used it to firmly fasten himself to one spot; or Cortez, who burned his ships behind him after landing in Mexico in 1519.]

Courage and Secrecy

The principle on which to manage an army is to set up one standard of courage which all must reach. How to make the best of both strong and weak is a question involving the proper use of ground.

Thus the skillful general conducts his army just as though he were leading a single man, willy-nilly, by the hand.

It is the business of a general to be quiet and thus ensure secrecy; upright and just, and thus maintain order. He must be able to mystify his officers and men by false reports and appearances, and thus keep them in total ignorance.

[Henderson, in his *Stonewall Jackson* says that "the infinite pains with which Jackson sought to conceal, even from his most trusted staff officers, his movements, his intentions, and his thoughts, a commander less thorough would have pronounced useless . . .]

By altering his arrangements and changing his plans, he keeps the enemy without definite knowledge of his movements. By shifting his camp and taking circuitous routes, he prevents the enemy from anticipating his purpose.

At the critical moment, the leader of an army should act like one who has climbed up a height and then kick away the ladder behind him. He carries his men deep into hostile territory before he shows his hand.

He burns his boats and breaks his cooking pots; like a shepherd driving a flock of sheep, he drives his men this way and that, and none knows whither he is going.

To muster his host and bring it into danger:—this may be termed the business of the general.

The different measures suited to the nine varieties of ground; the expediency of aggressive or defensive tactics; and the fundamental laws of human nature, are things that must most certainly be studied.

When invading hostile territory the general principle is that penetrating deeply brings cohesion; penetrating only a short way means dispersion.

In Strange Territory

When you leave your own country behind and take your army across neighbouring territory, you find yourself on critical ground. When there are means of communication on all four sides, the ground is one of intersecting highways.

When you penetrate deeply into a country, it is serious ground. When you penetrate but a little way, it is facile ground.

When you have the enemy's strongholds on your rear, and narrow passes in front, it is hemmed-in ground. When there is no place of refuge at all, it is desperate ground.

Therefore, on dispersive ground, I should inspire my men with unity of purpose. On facile ground, I should see that there is close connection between all parts of my army.

On contentious ground, I should hurry up my rear.

On open ground, I should keep a vigilant eye on my defenses. On ground of intersecting highways, I should consolidate my alliances.

On serious ground, I should try to ensure a continuous stream of supplies.

[Giles points out in his translation that this refers to forage and plunder, and not as one might expect, to unbroken communications.]

On difficult ground, I should keep pushing on along the road.

On hemmed-in ground, I should block any way of retreat. On desperate ground, I should proclaim to my soldiers the hopelessness of saving their lives. For it is the soldier's disposition to offer an obstinate resistance when surrounded, to fight hard when he cannot help himself and to obey promptly when he has fallen into danger.

We cannot enter into alliance with neighbouring princes until we are acquainted with their designs. We are not fit to lead an army on the march unless we are familiar with the face of the country—its mountains and forests, its pitfalls and precipices, its marshes and swamps. We shall be unable to turn natural advantages to account unless we make use of local guides.

Summary of Terrain

[This concludes what Sun Tzu has to say about "grounds" and the "variations" corresponding to them. "Reviewing the passages which bear on this important subject," Giles says, "we cannot fail to be struck by the desultory and unmethodical fashion in which it is treated." His elaboration of this is given here in substance.

[Sun Tzu begins abruptly to consider the variations of terrain in Chapter Eight before touching on terrain at all, but he only mentions five types of terrain, numbers 5, 7, 8 and 9 of the subsequent list, and one that is not included in it. In Chapter Nine he deals with a few varieties of grounds and then in Chapter Ten he sets forth six new types, with six variations. None of these is mentioned again. At last in Chapter Eleven, we come to the Nine Grounds *par excellence*, immediately followed by the variations, and then by considerable repetition. "Though it is impossible to account for the present state of Sun Tzu's text, a few suggestive facts may be brought into promience: Chapter Eight, according to the title, should deal with nine variations, whereas only five appear. It is an abnormally short chapter. Chapter Eleven is entitled The Nine Grounds. Several of these are defined twice over, besides which there are two distinct lists of the corresponding variations. The length of Chapter Eleven is disproportionate, being double that of any other except Nine. I do not propose to draw any inferences from these facts, beyond the general conclusion that Sun Tzu's work cannot have come down to us in the shape in which it left his hands: Chapter Eight is obviously defective and probably out of place, while Eleven seems to contain matter that has either been added by a later hand or ought to appear elsewhere.]

Principles of Importance

To be ignorant of any one of the following principles does not befit a warlike prince.

When a warlike prince attacks a powerful state, his generalship shows itself in preventing the concentration of the enemy's forces. He overawes his opponents, and their allies are prevented from joining against him.

Hence he does not strive to ally himself with all and sundry, nor does he foster the power of other states. He carries out his own secret designs, keeping his antagonists in awe. Thus he is able to capture their cities and overthrow their kingdoms.

Bestow rewards without regard to rule, issue orders without regard to previous arrangements and you will be able to handle a whole army.

Confront your soldiers with the deed itself; never let them know your design. When the outlook is bright, bring it before their eyes; but tell them nothing when the situation is gloomy.

Place your army in deadly peril and it will survive; plunge it into desperate straits and it will come off in safety.

For it is precisely when a force has fallen into harm's way that it is capable of striking a blow for victory.

Success in warfare is gained by carefully accommodating ourselves to the enemy's purpose.

[In other words, feign stupidity by appearing to yield and fall in with the enemy's wishes.]

Accommodate the Enemy

By persistently hanging on the enemy's flank, we shall succeed in the long run in killing the commander-in-chief. This is called ability to accomplish a thing by sheer cunning.

On the day that you assume your command, block the frontier passes, void the official passports and stop the passage of all emissaries.

Be stern in the council chamber, so that you may control the situation.

If the enemy leaves a door open, you must rush in.

Forestall your opponent by seizing what he holds dear, and subtly contrive to time his arrival on the ground.

Walk in the path defined by rule, [some authorities render this line as "Discard hard and fast rules" which is certainly more in keeping with Sun Tzu's philosophy of war] and accommodate yourself to the enemy until you can fight a decisive battle.

At first, then, exhibit the coyness of a maiden, until the enemy gives you an opening; afterwards emulate the rapidity of a running hare, and it will be too late for the enemy to oppose you.

⟦ TWELVE ⟧

Attack by Fire

SUN TZU said: There are five ways of attacking with fire. The first is to burn soldiers in their camp; the second is to burn stores; the third is to burn baggage-trains; the fourth is to burn arsenals and magazines; the fifth is to hurl dropping fire amongst the enemy.

In order to carry out an attack with fire, we must have means available; the material for raising fire should always be kept in readiness.

There is a proper season for making attacks with fire, and special days for starting a conflagration.

The proper season is when the weather is very dry; the special days are those when the moon is in the constellations of the Sieve, the Wall, the Wing or the Cross-bar; for these are all days of rising wind.

> [These correspond roughly to Sagittarius, the Archer, autumn; Taurus, the Bull, spring; Leo, the Lion, summer; and Aquarius, the Water Bearer, winter.]

In attacking with fire, one should be prepared to meet five possible developments:

When fire breaks out inside the enemy's camp, respond at once with an attack from without.

If there is an outbreak of fire, but the enemy's soldiers remain quiet, bide your time.

When the force of the flames has reached its height, follow it up with an attack, if that be practicable; if not stay where you are.

If it is possible to make an assault with fire from without, do not wait for it to break out within, but deliver your attack at the favorable moment.

When you start a fire, be to windward of it. Do not attack from the leeward.

A wind that rises in the daytime lasts long, but a night breeze soon fails.

Use of Fire and Water

In every army, the five developments connected with fire must be known, the movements of the stars calculated and watch kept for the proper days.

Hence those who use fire as an aid to the attack show intelligence; those who use water as an aid to the attack gain an accession of strength.

By means of water an enemy may be intercepted, but not robbed of all his belongings.

Unhappy is the fate of one who tries to win his battles and succeed in his attacks without cultivating the spirit of enterprise; for the result is waste of time and general stagnation.

Hence the saying: The enlightened ruler lays his plans well ahead; the good general cultivates his resources.

Move not unless you see an advantage; use not your troops unless there is something to be gained; fight not unless the position is critical.

No ruler should put troops into the field merely to gratify his own spleen; no general should fight a battle simply out of pique.

If it is to your advantage to make a forward move, make a forward move; if not, stay where you are.

Anger may in time change to gladness, vexation may be succeeded by content.

But a kingdom that has once been destroyed can never come again into being; nor can the dead ever be brought back to life.

Hence the enlightened ruler is heedful, and the good general full of caution. This is the way to keep a country at peace and an army intact.

[THIRTEEN]

Use of Spies

SUN TZU said: Raising a host of a hundred thousand men and marching them great distances entails heavy loss on the people and a drain on the resources of the state. The daily expenditure will amount to a thousand ounces of silver. There will be commotion at home and abroad, and men will drop down exhausted on the highways. As many as seven hundred thousand families will be impeded in their labor.

Hostile armies may face each other for years, striving for victory which is decided in a single day. This being so, to remain in ignorance of the enemy's condition simply because one grudges the outlay of a hundred ounces of silver in honours and emoluments, is the height of inhumanity.

One who acts thus is no leader of men, no present help to his sovereign, no master of victory.

Thus, what enables the wise sovereign and the good general to strike and conquer, and achieve things beyond the reach of ordinary men, is *foreknowledge.*

Now this foreknowledge cannot be elicited from spirits; it cannot be obtained inductively from experience, nor by any deductive calculation.

Classes of Spies

Knowledge of the enemy's dispositions can only

be obtained from other men. Hence the use of spies, of whom there are five classes: Local, inward, converted, doomed and surviving spies.

When these five kinds of spy are all at work, none can discover all of the ramifications of your secret spy system. This is called "divine manipulation of the threads." It is the sovereign's most precious faculty.

Local spying invaders employing the services of the inhabitants of a district.

Inward spies, making use of officials of the enemy.

Converted spies, getting hold of the enemy's spies and using them for our own purposes.

Doomed spies, doing certain things openly for purposes of deception, and allowing our own spies to know them and report them to the enemy.

[One Chinese commentator has what seems the best explanation of doomed spies. "We ostentatiously do things calculated to deceive our own spies; who must be led to believe that they have been unwittingly disclosed. Then, when these spies are captured in the enemy's lines, they will make an entirely false report, and the enemy will take measures accordingly, only to find that we do something quite different. The spies will thereupon be put to death."]

Surviving spies, those who bring back news from the enemy's camp.

Hence it is that with none in the whole army are more intimate relations to be maintained than with spies. None should be more liberally rewarded. In no other business should greater secrecy be preserved. Spies cannot be usefully employed without certain intuitive sagacity.

Methods of Espionage

They cannot be properly managed without benevolence and straightforwardness.

Without subtle ingenuity of mind, one cannot make certain of the truth of their reports. Be subtle, and use your spies for every kind of business.

If a secret piece of news is divulged by a spy before the time is ripe, he must be put to death together with the man to whom the secret was told.

Whether the object be to crush an army, to storm a city, or to assassinate an individual, it is always necessary to begin by finding out the names of the attendants, the aides-de-camp, the doorkeepers and sentries of the general in command. Our spies must be commissioned to ascertain these.

The enemy's spies who have come to spy on us must be sought out, tempted with bribes, led away and comfortably housed. Thus they will become converted spies and available for our service.

It is through the information brought by the converted spy that we are able to acquire and employ local and inward spies.

It is owing to his information, again, that we can cause the doomed spy to carry false tidings to the enemy. [Because the converted spy knows how the enemy can best be deceived].

Lastly, it is by his information that the surviving spy can be used on appointed occasions.

End and Aim of Spying

The end and aim of spying in all its five varieties is knowledge of the enemy; and this knowledge can only be derived, in the first instance, from the converted spy. Hence it is essential that the converted spy be treated with the utmost liberality.

Of old, the rise of the Yin dynasty [1766-1122 B.C.] was due to I Chih who had served under the Hsia [2205-1766 B.C.]. Likewise, the rise of the Chou dynasty [1122-256 B.C.] was due to Lü Ya who had served under the Yin.

Hence it is only the enlightened ruler and the wise general who will use the highest intelligence of the army for purposes of spying, and thereby they achieve great results. Spies are a most important element in war, because on them largely depends an army's ability to move.